Farm Poultry Management

by US Dept of Agriculture

with an introduction by Jackson Chambers

This work contains material that was originally published in 1977.

This publication is within the Public Domain and was originally published with Public Funding for the Public Benefit.

This edition is reprinted for educational purposes and in accordance with all applicable Federal Laws.

Introduction Copyright 2017 by Jackson Chambers

Self Reliance Books

Get more historic titles on animal and stock breeding, gardening and old fashioned skills by visiting us at:

http://selfreliancebooks.blogspot.com/

Introduction

I am pleased to present yet another title on Poultry.

The work is in the Public Domain and is re-printed here in accordance with Federal Laws.

As with all reprinted books of this age that are intended to perfectly reproduce the original edition, considerable pains and effort had to be undertaken to correct fading and sometimes outright damage to existing proofs of this title. At times, this task is quite monumental, requiring an almost total "rebuilding" of some pages from digital proofs of multiple copies. Despite this, imperfections still sometimes exist in the final proof and may detract from the visual appearance of the text.

I hope you enjoy reading this book as much as I enjoyed making it available to readers again.

Jackson Chambers

Contents

	Page		Page
Profitable Poultry Management	3	Brooding—Continued	
Size	3	Vaccination	21
Hatching versus market eggs	3	Rearing Started Pullets	22
Meat production	4	Confinement	22
Costs	4	Range	23
The Farm Laying Flock	4	Management of Layers	24
Egg-laying tests	5	Preparing the house	24
Source	6	Moving pullets	24
Getting started	6	Care of pullets	24
Housing	7	Lighting	25
Space	7	Culling	25
Building features	8	Marketing eggs	26
Equipment	9	High-quality eggs	26
Brooder house equipment	9	Care and handling of eggs	26
Growing house equipment	11	Hatching-Egg Production	28
Range equipment	11	Breeding flock management	29
Laying house equipment	11	Care of hatching eggs	30
Feed and Water	13	Meat Production	30
Water	14	Housing	31
Feed management	14	Brooding	31
Feeding replacement chicks	15	Caponizing	31
Feeding growing birds	16	Feeding meat-type birds	32
Feeding birds on range	16	Management	33
Feeding layers	16	Marketing poultry	33
Feeding breeders	18	Diseases, Parasites, and Pests	34
Brooding	19	Disease control	34
Time to brood	19	Mites, lice, and ticks	35
Preparing the house	19	Rats and mice	35
Operating the brooder	20	Predators	36
Management	20	Precautions	36
Cannibalism	21		

Washington, D.C.   Revised June 1977

For sale by the Superintendent of Documents, U.S. Government Printing Office
Washington, D.C. 20402
Stock No. 001-000-03643-2

Farm Poultry Management

Graham H. Purchase, ARS staff scientist [1]

Farm poultry flocks are kept principally for egg production. If farm flocks are obtained from bred-to-lay stocks and are properly managed, they can maintain a high rate of production throughout the year.

Specialty meat-type birds—capons and roasters—are particularly suited to farm production. Broilers usually cannot be grown on the farm as economically as they can be grown commercially.

PROFITABLE POULTRY MANAGEMENT

A farm flock can be profitable if the poultryman—

- Maintains a large enough flock so that he uses labor economically.
- Produces a high-quality product—market eggs, hatching eggs, or specialty meats.
- Starts with high-quality birds.
- Uses good feeds.
- Keeps the poultry house clean and dry.
- Employs sound management practices.
- Employs sound marketing practices.

Size

The trend is toward larger farm poultry units. There are many small flocks, however, that contribute to farm income.

The exact size of the poultry flock that is economically feasible is influenced by other farm enterprises, as well as by the geographical location and distance to markets. Guides to minimum flock sizes for efficient farm poultry production listed below.

A laying unit for a farm should have a minimum of 2,500 birds. A breeding (hatching egg) flock should have at least 2,500 pullets or hens; a capon or roaster flock, 2,000 birds; and a broiler flock, 6,000 birds.

Larger flocks usually are more economical.

Hatching Versus Market Eggs

Market eggs may be sold to wholesale or retail outlets. Fertile hatching eggs usually are sold directly to hatcheries.

[1] Beltsville Agricultural Research Center, Beltsville, Md. 20705.

Most farm breeding flocks produce hatching eggs under contract with a hatchery. The hatchery may specify the strain or cross of the breeding flock. It may provide pullet or cockerel chicks—or both—to be used as breeders.

Hatching eggs cost more to produce then market eggs. The added expenses include the costs of—

- Raising and keeping cockerels.
- Providing special (breeder) feeds.
- Blood testing birds for pollurum disease and fowl typhoid.
- Increased feed consumption, particularly of broiler parent stock.

To justify these costs of production, hatching eggs must sell at a premium over market eggs.

Meat Production

Before undertaking the production of meat chickens, a poultryman should determine how and where his birds will be marketed.

Broilers (young chickens for frying, broiling, or roasting) once were largely a byproduct of egg production. Today's commercial broilers are scientifically bred to produce meat efficiently. Commercial production is on a year-round basis; less than 3 percent of the Nation's total poultry meat production comes from farm flocks as a byproduct of egg production.

Roasters and capons are best suited for farm production. Roasters are 3- to 5-month-old chickens that weigh 5 pounds or more; capons are 6- to 7-month-old chickens that weigh 7 pounds or more. The per pound and per bird costs of raising these tender-meated, heavy-weight chickens are higher than the corresponding costs of raising broilers. Therefore, roasters and capons sell at a higher price per pound.

The market for roasters and capons is relatively undeveloped, but has a good potential. Farm production of these specialty products should increase significantly in areas where consumer demand can be built up.

Costs

Feed is a major expense in poultry production. It amounts to about 55 percent of the total cost of producing market eggs, 65 percent of the cost of producing meat-type birds, and 55 percent of the cost of producing hatching eggs.

The age and breeding of chickens determine their purchase price. Highest purchase prices are charged for ready-to-lay pullets. Day-old sexed pullet chicks sell for 25 to 30 percent of the cost of ready-to-lay pullets of the same breeding.

Normal expenses of a farm flock—in addition to feed and birds—include housing, equipment, vaccines and drugs, fuel, electricity, labor, taxes, interest, and depreciation.

Accurate records are essential in figuring costs and determining efficiency. Records also can be used to improve operations.

THE FARM LAYING FLOCK

Start the laying flock with top-quality, healthy birds that have been developed for a high rate of production throughout the year. With

White Leghorn (top), a light-weight breed that lays white eggs, is the most popular bird for farm laying flocks. Rhode Island Red (center) and New Hampshire (bottom) are medium-weight breeds that lay brown eggs.

proper management, bred-to-lay birds produce eggs of superior size and quality.

Before buying birds, decide whether you want white or brown eggs. The color of the eggshell does not affect food value, but it does influence the market price. In most sections of the United States, white-shelled eggs sell for slightly more than brown-shelled eggs. In New England, where brown eggs are preferred, the situation is reversed.

Birds that weigh 4 to 4½ pounds at maturity are considered light-weight breeds. They usually lay white eggs. White Leghorn first-generation strain crosses and crosses of inbred lines are the most popular light-weight varieties for laying flocks.

Medium-weight breeds usually lay brown eggs. Popular medium-weight egg-laying breeds are New Hampshires, Plymouth Rocks, Rhode Island Reds, and first-generation egg-production crossbreds.

Egg-Laying Tests

Random sample egg-production tests are conducted in many States. These tests are designed to provide a reliable guide to the performance of laying stocks offered for sale by breeders and hatcherymen. The tests provide information on all important economic traits of poultry.

The U.S. Department of Agriculture annually publishes a report on the egg-production tests. Copies of the current publication, which contains records of all stocks entered in performance tests in the United States and Canada, may be obtained by writing to the Animal Improvement Programs Laboratory, Agricultural Research Service, U.S. Department of Agriculture, Agricultural Research Center, Beltsville, Maryland 20705.

Source

Bred-to-lay birds are available as chicks from hatcheries and breeding farms, and as started pullets from these sources and from specialized pullet growers. Stocks that have ranked high in random-sample tests often are sold by regional distributors.

Large commercial hatcheries have taken over most of the hatching formerly done on the farm or at small community hatcheries. In most cases, farm hatching is no longer profitable. Commercial hatcheries sell day-old chicks of high quality for less than it would cost to produce chicks in a farm incubator.

Get your stock from the nearest source that offers birds with the traits you want. The shorter the distance your birds must travel to the farm, the fewer your losses.

It is a good idea to investigate the local reputation—as well as the breeding, sanitation, and management practices—of the hatchery or grower before placing an order for chickens.

To find out more about the kind of birds suited to your needs, talk with your county agricultural agent, a poultry specialist, or successful poultrymen in your area.

Getting Started

Begin planning for the flock at least 6 months before you want the chickens on your farm. Select the age of birds that are best suited to your production timetable, your houses and equipment, and local conditions. Order chicks at least 4 weeks in advance; order started pullets at least 3 months before you need them; order ready-to-lay pullets at least 6 months ahead.

Plan to keep only one age of bird in a flock at one time. If you must have birds of several ages on the farm, keep each age group separated to help reduce diseases.

Baby Chicks

Baby chicks usually leave the hatchery when they are 1 day old. They normally require heat during the early weeks. They sometimes are vaccinated, dubbed, sexed, and debeaked at the hatchery.

Straight-run chicks cost less than other ages and types of live birds. Straight-run birds are boxed at random as they come from the incubator; normally, about half the chicks are pullets and half are cockerels. When space and labor are available on the farm, cockerels of medium-weight breeds may be grown as meat-type birds at the same time that pullets are grown as layers. Sometimes, cockerels of superior strains are raised as breeding stock.

Sexed chicks are sorted after they leave the incubator into lots of pullets and cockerels. Sexed pullets of light-weight breeds cost more than twice as much as straight-run chicks; sexed pullets of medium-weight breeds—while usually not as expensive as light-weight breeds—cost more than straight-run chicks.

Started birds

Sales of started birds are increasing. These chickens, which have been brooded and no longer need supplemental heat, require less equipment and less care than younger birds.

In many sections of the country, started pullets are available from growers who specialize in their production. There should be an understanding between the buyer and the grower concerning the strain or cross of the started pullets, the type and number of vaccinations, the kind of feed, and the disease history of the birds.

Started pullets usually are sold between 6 and 8 weeks of age. Individual lots may vary with seasonal needs for heat. Their high purchase price reflects the costs involved in getting the birds through brooding.

Ready-to-lay pullets are sold at 16 to 20 weeks of age. They cost more than other classes of poultry of similar age because of the feed, care, culling, and management involved in developing them for egg production.

Number

Plan to fill your laying house without overcrowding. In determining the number of birds you need, allow for normal losses from diseases, natural causes, and culling.

For each 100 layers you want in your flock, start with—
- 220 straight-run chicks (day old).
- 110 sexed pullets (day old).
- 105 started pullets (6 weeks old).
- 100 ready-to-lay pullets.

HOUSING

Birds of different ages always should be housed separately.

The trend is toward raising poultry entirely in confinement. Two systems may be used:
- Floor housing (sometimes called litter or "loose" housing).
- Cages.

Floor housing allows birds to move freely on the floor of the poultry house, or inside pens that divide large floor areas into manageable units. This system may be adapted to brooding, growing, and keeping any type of poultry flock—layers, breeders, broilers, capons, or roasters. It is particularly suited for farm flocks.

Cage raising methods are best adapted to situations where land is limited, or to areas in the southern United States where temperatures rarely go below freezing. Laying hens in cooler areas can be housed in cages if—
- The house is well insulated.
- Ventilation is controlled.
- The house is filled to capacity so that the hens' bodies provide the necessary heat to maintain a minimum temperature range of 40° to 55° F.

Space

Space requirements per bird in a floor housing system are the same for all types of chicks up to 10 weeks. In general, chicks less than 6 weeks old need ½ square foot per bird; chicks from 6 to 10 weeks need 1 square foot per bird; growing pullets from 10 to 20 weeks need 1½ to 2 square feet per bird. Layers need

2 to 3 or more square feet per bird, the exact space depending on the body size of the bird and the temperature. Pullets need about ½ square foot more space per bird when the temperature is above 80° F.

As a general rule, large flocks require less space per bird than small flocks.

Through efficient management and use of automatic equipment, some poultrymen have reduced space allowances per layer without increasing losses. When space per layer is reduced, the layers should be debeaked. More care is required in operating the house to maintain egg production if space is limited.

Some poultrymen with small flocks make use of porches or yards to provide additional space per bird.

A 12- by 16-foot brooder house will accommodate 200 pullet chicks or 200 straight-run chicks to 8 weeks. A brooding-growing house should be 2 to 3 times larger than a brooding house that holds the same number of birds. A 36- by 140-foot house will accommodate 3,300 small-breed pullets to laying age, 2,500 pullets and cockerels to maturity, or 6,000 broilers to market age.

A portable range shelter 8 by 9 feet will house 100 birds from the time they go onto the range until they reach maturity.

Building Features

A poultry house should provide clean, dry, comfortable quarters for birds throughout the year. An interior temperature range of 45° to 80° F is satisfactory for economy of egg production.

Moisture is a common problem in poultry houses. Fresh air should be circulated, but the house should be free of drafts.

Pole-type houses are economical to build, maintain, and clean. They are becoming increasingly popular for all types of flocks—replacement chicks, growing pullets, layers, breeders, and meat-type birds. Pressure-treated poles are set into the ground to support the roof and walls of a single-story structure. Pole-type houses often have roofs made of metal or other waterproofed materials. Walls may be made of poultry netting or conventional building materials.

Other types of poultry buildings include open, open-front, and insulated houses. Prefabricated poultry houses of several types are available.

State agricultural colleges, in cooperation with the U.S. Department of Agriculture, have developed poultry-house plans for specific States and regions. Plans may be ordered through county agents or from agricultural colleges. (Plans are *not* available from the USDA.)

All poultry buildings—whether new or remodeled—should contain—

• Floors which can be cleaned and disinfected easily.

• Walls which can be washed easily.

• A ventilation system. Built-in gravity systems, adjustable roof ridge ventilators, and fans are often used.

• Fireproof and vermin proof insulation.

• Water piped into the house. The water system should be adequately protected against freezing.

Pole-type laying house.

- Electricity, for artificial lights to provide a uniform light "day" in the poultry house, and to operate equipment, as needed.

Auxiliary rooms—such as those for egg, feed, or equipment storage—should be centrally located.

Some States require incinerators or disposal pits on all poultry farms, as a disease-control measure.

Incinerators or disposal pits should be placed where they are convenient. A disposal pit 4 feet square and 6 feet deep is adequate for a flock with 1,500 layers or 10,000 broilers. It should be constructed above the water table, at least 100 feet from the water supply, with drainage in the opposite direction. The cover should be tight fitting. Local public health officials should be consulted in installing a disposal pit.

EQUIPMENT

Select equipment that is safe for the birds, convenient to use, and easy to keep clean. Whenever possible, equipment should be installed in a way that allows removal when the house is cleaned.

Use of automatic or semiautomatic equipment saves hours of labor. Mechanical feeders, waterers, and—occasionally—pit cleaners are used for farm-sized flocks. Feed and litter carriers that operate on tracks are adapted to use in long houses. Bulk feed requires special storage bins.

Brooder House Equipment

Brooders

To give day-old chicks a proper start, the brooder must provide a temperature of about 95° F in winter and 90° the rest of the year.

Some types of brooders heat the entire room or house. Other types warm the area under or near the hover, while the rest of the room remains relatively cool. Chicks feather better when they have a cool place to exercise.

Chicks need enough space under the brooder so that they can keep warm without crowding, piling up, or smothering. Under normal conditions, each replacement chick needs about 6 or 7 square inches of brooder space. Manufacturers often overrate the capacity of their brooders. For example, a 96-inch hover—rated as a 1,000-chick brooder—will satisfactorily brood 600 replacement or 850 broiler chicks under winter conditions.

Brooders may be operated by hot water or hot air. The fuel may be electricity, oil, gas, or coal.

In selecting brooders, consider the fuels available in your locality and the amount of heat needed during the seasons in which you intend to brood. Choose brooders that can be raised and lowered easily, are easy to clean, have reliable thermostats, and are equipped with thermometers.

Electric and gas brooders require minimum care.

In cold weather, electric brooders should be used only in well-insulated houses. If a house is not insulated, these brooders may not give off enough heat during the winter to keep chicks warm and the litter dry. Auxiliary heat may be needed during the winter with electric brooders.

Gas and oil-burning brooders vary widely in heat output. In cold weather, wet litter may be a special problem with gas brooders that do not have flues for venting combustion gases to the outside.

Coal-burning brooders require more labor than other systems, because they must be refueled and cleaned frequently. They keep the entire house warm and the litter relatively dry.

Fires in poultry houses often start from leaky oil brooders. Coal brooders also may be fire hazards if not properly operated. An asbestos sheet, or other fire-resistant material, may be placed under brooders to minimize the danger from fires.

Lights

A 7½- or 15-watt light under the hover will attract young chicks to heat. Attraction lights are not normally used after the first 2 weeks of brooding.

Feeders

Hanging tube-type feeders are rapidly replacing trough-type feeders, because there is less labor and there is less bruising of chicks with the hanging feeders.

Allow three hanging feeders—each 15 inches in diameter with a 25-pound capacity—per 100 chicks.

Hanging feeders.

If trough feeders are available, they may be used for a farm flock. Allow two 4-foot trough feed hoppers, open on both sides, or 200 linear inches of hopper space for each 100 chicks at 3 weeks of age.

When chicks are 7 weeks old, provide three 4-foot feed hoppers, or 300 linear inches, per 100 birds. Provide additional feeders when the temperature is above 80° F.

If automatic feeders are used, follow the manufacturer's directions carefully.

Waterers

Use two 1-gallon water fountains for each 100 chicks during the first 2 weeks. Allow 40 linear inches of water trough per 100 birds at 3 weeks, and 50 linear inches from the time the birds are 7 weeks old until the end of brooding. Provide additional waterers when the temperature is above 80° F.

Waterers should be automatic or float-controlled. Hanging waterers are preferable to the stationary type.

Follow the manufacturer's recommendations concerning space allowances for automatic water systems.

Other equipment

Roosts should be provided in the brooding house if roosts are to be used in the laying house. (See Laying House Equipment.)

Growing House Equipment

Feeders, waterers, and electric lights are standard equipment for growing houses. Requirements for feeders and waterers are the same as those for brooding houses.

If roosts are to be used in the laying house, they should be provided for pullets in the growing house.

Range Equipment

On range, provide three 6-foot feeders open on both sides for each 100 birds, or 4 inches of feeder space per bird. To supply water on range, use one 8-gallon gravity-flow waterer, or 8 linear feet of a trough-type waterer, or a fountain with an equivalent capacity for each 100 birds.

Feed hoppers and waterers for range should be close to the shelter. Range equipment may include a protective canopy or sunshade.

Many range shelters have no interior equipment. Some have one feed trough and one fountain inside the shelter for use in bad weather.

Roosts and nests should be installed in range shelters before pullets begin to lay. Place roosts at floor level.

Laying House Equipment

Nests

Well-designed nests can reduce the time needed to care for the laying flock and the eggs. Nests may be metal or wood; they may have roll-away floors with egg trays, or other arrangements for convenience in collecting eggs. The interiors should be dark.

Nests may be placed in the middle or along the walls inside of the building. They sometimes are arranged in a double deck.

Community nests accommodate several layers at one time. Allow one nest 2 feet wide by 14 feet long, with an entrance 8 inches square, for each 140 hens; or provide 1 square foot for nesting space for each 5 hens.

Individual nests are just large enough to hold one hen. In figuring flock needs, provide one individual nest for each four birds. Usually, an individual nest is 10 to 12 inches wide, 12 to 14 inches high, and about 12 inches deep. A perch below the entrance will help keep the nest clean.

Roosts

Roosts should always be used for growing pullets that are later to be maintained as layers in houses with roosts.

Roosts should be made of 2-inch stock, with rounded or beveled upper edges. Leghorns and other small breeds require 8 inches of roost space per bird; large breeds require about 10 inches of space per bird. Place roosts 13 to 15 inches apart. Normally, roosts are placed above dropping pits.

Roosts, if used, should be built on a slant from the floor. The back of the roosts should be 24 inches from the floor, with 1-inch mesh wire beneath them.

Multiple-deck roosts sometimes are built over dropping pits for use with flocks of more than 1,000 birds. Housing needs are reduced by ½ square foot of floor space per bird with multiple-deck roosts.

Dropping pits

A dropping pit is designed to hold droppings for several months. For ease in handling, the floor over the pit may be made into 6- by 6-foot panels. There is a real saving in labor from the use of pits. However, pits harbor rats and are a breeding place for flies.

Feeders and waterers often are placed over pits.

Feeders

Allow six hanging feeders—each 15 inches in diameter with a 35- to 50-pound capacity—per 100 medium-weight layers. If hanging feeders are not available, provide at least 40 feet of feeder space (four 5-foot through feeders with both sides open, or the equivalent) for each 100 laying hens.

Automatic feeders vary widely in capacity. Consult the manufacturer's literature or a specialist before installing such equipment.

All feeders should be placed within 10 feet of a waterer.

Hoppers are needed for insoluble grit and calcium supplements, if these are not included in the feed. For each 100 hens, provide a 12-inch granite grit hopper box and a 12-inch hopper for oystershell or limestone grit.

Waterers

Provide an 8-foot automatic hanging waterer, open on both sides, or an equivalent 16 feet of watering space, for each 200 pullets in the laying house. Increase watering space

25 percent when temperatures go above 80° F.

If watering devices are placed on roosts over dropping pits, the amount of wet litter in the laying house will be minimized.

Water systems may be automatically controlled, or they may flow continuously. Adequate drainage should be provided.

Water requirements vary with the type of waterer and the season. An automatic system uses 6 to 8 gallons of water daily for each 100 layers.

A reserve water supply is helpful in disease outbreaks or times of disaster. One or more clean oil drums can be used for water storage.

Lights

Before installing light fixtures, be sure that the electrical wiring is adequate and that it meets all safety and local code requirements. If you need assistance, contact your local power company.

Automatic switches to control lights are inexpensive and easy to install. An automatic time clock may be set to turn lights on and off at desired times. Automatic dimming devices are recommended when evening lights are used.

Allow one ceiling light for each 200 square feet of floor space. Adjust lights to illuminate the entire floor and roosting areas.

For daytime, morning, or evening lighting, use 60-watt incandescent light bulbs. Shallow-dome (aluminum pie-plate) reflectors or bulbs with built-in reflectors improve the distribution of light within the poultry house.

FEED AND WATER

Poultry feeds should provide proteins, carbohydrates, fats, minerals, vitamins, and added growth factors in proper balance. Incomplete or unbalanced rations often result in nutritional diseases.

All types of poultry feeds are available as manufactured—or commercial formula—feeds. Always follow manufacturer's instructions for using commercial feed; provide supplements as directed.

Mixing poultry feeds on the farm is no longer economical, unless home-grown grains make up a greater part of the mash. Custom mixing, which combines a concentrate with home-grown grain, may reduce feed costs.

In selecting a feed ration, consider which feed will give the best results at the lowest practical cost. If the formula is open, consider the kind and amount of ingredients, the percentage of protein, and the energy content, as well as the cost. With a closed-formula feed, consider the reputation of the feed itself and the company that manufactures it, in addition to its cost.

Feed formulas are given in this publication as examples of the types of feeds suitable for farm flocks. Each sample ration includes the feed constituents needed to satisfy the nutrient requirements of poultry at a designated period of growth. It should not be assumed that these are the only formulas available, or that they are the best and most economical under all conditions at all times in all sections of the United States.

In addition to regular (ground) mash, feed is available as pellets and crumbles. The processing of feed into pellet or crumble form increases the cost over the mash form. It also increases palatability and usually reduces feed waste. Sometimes, the nutritive value of the feed is increased during processing.

Certain drugs have been approved for use in poultry feeds. These drugs may help prevent or reduce disease outbreaks, enable birds to meet stress conditions, and speed the growth of chicks.

Water

Keep clean fresh water before birds at all times. Provide at least 1½ gallons per day for each 100 chicks for the first 3 weeks. Increase the supply to 3 gallons per day for each 100 chicks at 6 weeks, to 5 gallons at 10 weeks, and to 6 gallons at 12 or 13 weeks. Supply 6 to 8 gallons of water daily for each 100 layers. Provide additional water when temperatures are above 80° F; for example, at temperatures above this level, 100 layers require 9 gallons of water per day.

When a medicated feed is used, follow the manufacturer's directions exactly. Directions on the label or feed tag are based on U.S. Food and Drug Administration regulations governing the use of a medicated feed, its withdrawal, and replacement with a nonmedicated feed.

Thoroughly clean water fountains or troughs once a day or more often, if necessary. Refill with a clean, fresh supply of water. Be careful not to spill or empty water on litter.

Feed Management

To maintain healthy birds, keep fresh feed available at all times.

Limit the amount of feed in feeders to the extent necessary to avoid waste. It is a good practice to fill hanging feeds only three-fourths full, and trough feeders only two-thirds full. For efficient feeding, keep the lip of the feeder pan in a hanging tube-type feeder at the level of the birds' backs.

Fill nonautomatic trough feeders in the early morning, and during the day whenever feed supplies get low. If leftover feed is not clean and palatable, remove it before refilling feeders. Never put moldy or contaminated feed in feeders. Clean feeders as needed.

Keep a close check on birds' weight and their feed consumption. A drop in feed intake usually is the first indication of trouble—a disease outbreak, molt, stress, or poor management. If the reason for the drop in feed consumption is not readily apparent, consult a poultry specialist.

Keep feed as fresh as possible. Order for frequent delivery—if possible, every 2 weeks.

Store feed carefully, in a dry, rat- and mouse-proof place, where it will not be subject to damage from moisture or losses from rodents.

If the laying flock contains more than 2,000 birds, handle poultry feed in bulk.

Feeding Replacement Chicks

Starter rations for replacement chicks should contain at least 20 percent of protein, the muscle-building nutrient.

A sample of a complete—or all-mash—starter feed formula is given in table 1. Like other sample rations given in tables 1, 2, and 3, this contains the nutrient allowances for poultry recommended by the National Research Council.

Always buy complete starter feeds for chicks. Give chicks starter feeds (without grain supplement) until they are 6 weeks old. Supply chick-sized granite grit in separate hoppers.

Allow 40 pounds of starter mash per 100 chicks for the first 2 weeks, and 250 to 300 pounds per 100 chicks from 2 to 6 weeks. Broiler replacements require more than small-breed replacements. Increase the size of granite grit as the birds mature.

Provide feed and water just as soon as chicks are placed in the brooder house. Spread mash on clean, cut-down chick boxes and lids to encourage chicks to start eating.

TABLE 1.—*Complete feeds for laying flock replacements*

Ingredient	Starter	Grower	Layer	Breeder
	Percent	Percent	Percent	Percent
Ground yellow corn	25.0	25.0	22.0	22.3
Ground oats or barley	10.0	10.0	10.0	10.0
Ground corn, wheat, barley, or grain sorghum	16.8	33.0	34.0	29.0
Alfalfa meal (17 percent protein)[1]	5.0	5.0		5.0
Soybean oil meal (44 percent protein)	23.0	16.0	13.0	12.0
Cottonseed meal, peanut meal, corn gluten meal, or soybean oil meal	10.0	5.0	5.0	5.0
Fish meal (60 percent protein)	2.0		2.0	2.0
Riboflavin supplement (20 micrograms riboflavin per gram)		2.0	4.0	5.0
Dicalcium phosphate (18 percent phosphorus)	1.7	2.5	2.5	3.2
Ground limestone or oystershell	1.0	1.0	7.0	6.0
Manganized salt[2]		.5	.5	.5
Coccidiostat	([3])	([3])		
Vitamin A	([4])		([5])	([6])
Vitamin D_3	([7])	([8])	([7])	([7])
Vitamin B_{12}				([9])
Total	100.0	100.0	100.0	100.0

[1] Meal with guaranteed vitamin A content of 100,000 International Units per pound.
[2] To supply 25 milligrams of manganese per pound of feed. Prepare by mixing 100 pounds of dairy or table salt with 3 pounds of technical anhydrous manganous sulfate.
[3] Give in the form and at the level recommended by the manufacturer.
[4] 3,000 International Units of stabilized vitamin A (dry form) per pound of feed.
[5] 4,000 International Units of stabilized vitamin A (dry form) per pound.
[6] 5,000 International Units of stabilized vitamin A (dry form) per pound.
[7] 500 International Chick Units (ICU) of stabilized vitamin D_3 (dry form) per pound.
[8] 400 International Chick Units of stabilized vitamin D_3 (dry form) per pound.
[9] 2 micrograms of vitamin B_{12} per pound.

Allow one box or lid for each 100 chicks.

If hanging feeders are used, place feeder pans on the floor about the fifth day. Raise the feeder and the plan gradually.

Chicks usually can learn to use hanging or automatic feeders and waterers by the time they are 2 weeks old. Allow a gradual change to automatic feeders and waterers.

Feeding Growing Birds

When chicks are 6 weeks old, change to a growing ration. Table 1 lists a sample of a complete growing ration; table 2, a sample ration to be fed with grain.

Pullets may begin to receive grain as soon as they start eating growing mash. Corn, wheat, barley, oats, millet, grain sorghum, or combinations of these may be used.

Begin with 10 pounds of grain for each 100 pounds of mash. Increase grain until pullets are getting equal parts of mash and grain. Put grain and mash in separate hoppers.

When pullets are 18 to 20 weeks old, gradually withdraw the growing mash and replace it with laying mash over a 2-week period.

Feeding Birds on Range

Range cannot provide a complete diet for birds. Pullets that get the green feed of the range need the additional nutrients of a growing ration. A good ration for range-reared birds is listed in table 2. Mash or pellets usually are fed in one hopper and grain is fed in another.

Some poultrymen use pellets for range feeding, because the larger particles are less subject to blowing out of feeders.

Feeding Layers

The best ration for layers is the one that will produce a dozen high-quality eggs at the least cost for feed. The actual cost of the feed that a layer eats in producing a dozen eggs—not the price per pound of feed—determines the economy of the ration.

Laying flocks need high-quality feed if they are to reach and maintain their maximum egg production. The amount of feed (in pounds) required to produce a dozen eggs decreases rapidly as egg production increases.

Feed consumption per bird varies primarily with egg production and body size. It also is influenced by the health of the bird and by the environment, especially temperature.

A hundred light-weight layers eat about 24 pounds of feed a day, while an equal number of heavy-weight birds eat about 30 pounds daily. Normally, a mature Leghorn or other light-weight bird eats 85 to 90 pounds of feed a year. A bird of a heavier breed eats 95 to 115 pounds. In addition, a layer consumes 2 to 5 pounds of oystershell and 1 pound of grit per year.

Complete feeds

Some poultrymen feed a complete laying feed that contains 15 or 16 percent of protein. Such feeds usually are more expensive than mash-and-grain rations. They normally require less labor, because no grain has to be handled. Complete feeds

TABLE 2.—*Feeds to be fed with grain for growing, laying, and breeding flocks*

Ingredient	Grower[1]	Layer[2]	Breeder[2]
	Percent	*Percent*	*Percent*
Ground yellow corn	25.0	23.0	24.0
Ground oats or barley	10.0	10.0	10.0
Ground corn wheat barley or grain sorghum	9.5	10.0	
Alfalfa meal (17 percent protein)[3]	5.0		5.0
Soybean oil meal (44 percent protein)	27.0	25.0	24.0
Cottonseed meal, peanut meal, corn gluten meal, or soybean oil meal	10.0	10.0	10.0
Fishmeal (60 percent protein)		4.0	4.0
Riboflavin supplement (20 micrograms riboflavin per gram)	6.0	7.0	12.0
Dicalcium phosphate (18 percent phosphorus)	4.5	5.0	6.0
Ground limestone or oystershell	2.0	5.0	4.0
Manganized salt[4]	1.0	1.0	1.0
Coccidiostat	([5])		
Vitamin A	([6])	([7])	([8])
Vitamin D₃	([9])	([10])	([11])
Vitamin B₁₂			([12])
Total	100.0	100.0	100.0

[1] Begin feeding 6-week-old birds 1 part of grain to 10 parts of mash; increase grain supplies until pullets receive equal parts of mash and grain.
[2] Use equal weights of mash and grain and supply limestone grit or oystershell in hoppers.
[3] Meal with guaranteed vitamin A content of 100,000 International Units per pound.
[4] To supply 25 milligrams of manganese per pound of feed. Prepare by mixing 100 pounds of dairy or table salt with 3 pounds of technical anhydrous manganous sulfate.
[5] Give in the form and at the level recommended by the manufacturer.
[6] 5,000 International Units of stabilized vitamin A (dry form) per pound of feed.
[7] 7,000 International Units of stabilized vitamin A (dry form) per pound.
[8] 8,000 International Units of stabilized vitamin A (dry form) per pound.
[9] 700 International Chick Units of stabilized vitamin D₃ (dry form) per pound.
[10] 800 International Chick Units of stabilized vitamin D₃ (dry form) per pound.
[11] 1,000 International Chick Units of stabilized vitamin D₃ (dry form) per pound.
[12] 4 micrograms of vitamin B₁₂ per pound.

are well suited for mechanical feeders. Table 1 (p. 15) includes an example of a complete laying feed.

Mash-and-grain

Most farm laying flocks are fed rations which contain equal parts of mash and grain. The mash generally has a protein content of 20 to 22 percent so that when it is fed with grain, the combination provides 15 or 16 percent of protein. A sample mash-and-grain ration is listed in table 2.

Although grain usually is fed in hoppers, it may be scattered in litter. Hens scratching for the grain keep the litter stirred up, which reduces moisture. This results in drier floors and fewer dirty eggs.

Provide extra grain in cold weather.

Supplements

Insoluble grit should supplement a mash-and-grain ration. (It is not used with all-mash feeds.) Grit may be fed in grit boxes or spread on top of mash in feeders.

Laying hens have high calcium requirements, because eggshells are largely calcium carbonate. Calcium should always be available. It may be fed as oystershell or limestone, in boxes or mixed with mash. Consumption of oystershell or limestone grit varies greatly.

Pellets and crumbles

Pellets and crumbles are used to stimulate appetites of birds suffering from diseases, effects of hot weather, or other stresses. These processed feeds have largely replaced wet mash for this purpose.

In feeding pellets or crumbles, add 3 to 5 pounds to the top of regular mash once a day for each 100 layers.

Feeding Breeders

Because breeder mash is more expensive than laying mash, use it only for breeding flocks. Start the birds on breeder mash about 1 month before hatching eggs are to be saved.

Breeder hens require greater amounts of vitamins A, D, and B_{12}, riboflavin, pantothenic acid, niacin, and manganese than laying flocks. Rations with these added ingredients in the right proportions give high hatchability and good development of chicks.

Breeding flocks may be fed complete rations, or equal parts of mash and grain. The complete ration

Chicks may be started on mash in box tops.

should contain at least 15 percent of protein. Sample breeder rations are listed in tables 1 and 2.

Although the rations differ, general recommendations for feeding laying flocks also apply to feeding breeding flocks.

If a mash-and-grain ration is used, give the breeding flock grain in the late afternoon or in the evening. Some poultrymen scatter about one-third of the day's grain in litter in the morning to stimulate activity and to help keep the litter dry.

Older hens may be fed a supplement of breeder pellets when necessary to stimulate their appetites.

BROODING

Chicks should be brooded in an environment that fosters normal, economical growth and good health.

Time to Brood

Adjust the brooding schedule to the flock you intend to replace. When the flock's egg production drops to an unprofitable level, the replacement pullets should be ready to lay or should already be in the early weeks of egg production. To achieve a smooth change in flocks, brooding must take place 5 to 5½ months before the replacements are needed.

Preparing the House

Prepare the brooder house several days before the new flock arrives on the farm. Advance preparations should include—
- Checking the house for potential trouble spots—such as leaks in the roof, leaky water spigots, holes in sidewalls, and excessive cracks in doors and windows.
- Making necessary repairs.
- Thorough cleaning and disinfection.
- Putting down fresh litter.
- Testing each brooder by operating it for at least 2 days. This includes making certain all thermostats are working properly and all thermometers are whole and in place.

Do not use old litter in brooding replacement chicks. The risks of disease are too great.

Litter may be any absorbent material that reduces the moisture in the poultry house and serves as an insulating material in cold weather. Built-up litter is more popular than the system that uses a thin layer of litter and requires frequent cleaning. The usual method of management is to put down clean, fresh litter before the flock is housed, and then to add new litter on top of it as needed. During cold weather, litter may reach a depth of 8 to 10 inches. Less litter is needed in hot weather.

The litter should be clean, mold free, and dry, but not dusty. Some poultrymen select litter from a material that is plentiful on the farm, such as wood shavings, sawdust, chopped straw, or corncobs. Others use a commercial litter, such as peat moss or sugarcane pulp (bagasse).

Place a 3- to 4-inch layer of litter on a base of clean, dry sand or directly on the clean floor of the brooder house several days before chicks arrive. Keep litter as dry as possible. Whenever necessary, stir the litter to keep it from packing. As wet spots develop, remove the wet and "caked over" litter and add new, dry material. If litter continues to be wet, apply 10 to 15 pounds of hydrated lime per 100 feet of floor space and stir.

Operating the Brooder

Start the brooders at least 2 days before you expect the chicks to arrive. Set the temperature at about 95° F in old weather and 90° in hot weather.

Make sure the temperature is adequate before placing the chicks under the hover. The room temperature for day-old chicks in a cold-room system should be maintained at a minimum of 65° F; where an economical source of fuel or a well-insulated house is available, it is desirable to maintain a temperature of about 75°.

Place the chicks under the hovers as soon as possible after they arrive. Keep chicks comfortable. Their actions provide a good guide to their temperature. Chicks crowd together near heat when they are too cold, and they pant and gasp (often at the outer edge of the hover guard) when overheated.

Follow the brooder manufacturer's recommendations on temperatures for operating the brooder. Normally, temperatures are reduced about 5° weekly until they reach 75°.

Check the temperature under the hover twice daily during the first week. Continue to check it twice a day as long as the chicks need heat. Temperatures should be taken about 3 inches above the floor and about 3 inches inside the outer edge of the hover.

With individual brooder stoves, use a hover guard of corrugated cardboard about 12 inches high to keep chicks near the heat. As soon as chicks are under the hover in cold weather, place the guard on the floor 2 feet from the outer edge of the brooder. Move the guard farther away from heat every day, and remove it after about a week, or when it is no longer needed. With debeaked chicks, use wire guard.

With infrared lamps or other systems that do not use a hover guard, block the corners of the house with cardboard or wire to prevent chicks from crowding into them and smothering.

Provide heat until chicks are well feathered. Birds are more likely to develop respiratory troubles if heat is removed too early. In late winter and early spring, heat may be continued for 8 weeks; in late spring and summer, for 4 to 6 weeks.

Do not crowd chicks. Birds may pile up or smother if they do not have enough space, or if they are frightened.

For summer brooding, protect chicks against temperatures above 95°. Keep them comfortable.

Management

If layers are to have roosts at maturity, chicks should have roosts in the brooder house. There usually is less crowding of birds and fewer deaths from piling up in houses with roosts. If roosts are used, chicks should be taught to roost as soon as they can get along without the hover protection.

Allow chicks enough space to grow properly. For the first 10 weeks, provide each bird 1 square foot of floor space.

If straight-run chicks are brooded, separate the cockerels from the pullets when the flock is 6 weeks old. This allows more floor space for pullets left in the house. The cockerels may be finished as broilers or fryers.

Cannibalism

Toe-, feather-, or body-picking may occur in confined flocks, even when birds are properly managed and well fed.

Toe picking may be the first sign of cannibalism. Remove injured chicks immediately and paint their toes with a stop-pick preparation. Allow injuries to heal before returning chicks to the flock.

Debeaking may be used either to halt picking or to prevent it. Some poultrymen debeak routinely; others practice it only when cannibalism appears. Debeaking may be done at any age. If previous flocks have engaged in cannibalism, it is a good idea to order chicks debeaked at the hatchery. Early debeaking may last only 10 weeks.

When pullets are debeaked at 16 to 20 weeks (and at least 2 weeks before they are moved into the laying house), cannibalism normally is prevented for the rest of the birds' lives.

Debeaking does not damage the health or reduce the vigor of birds.

Vaccination

Effective vaccines have been developed for four major respiratory diseases: Newcastle disease, bronchitis, fowlpox, laryngotracheitis.

Plan a vaccination schedule to cover the flock from the time the chicks arrive on the farm until they complete a year in the laying house. Base the vaccination schedule on the needs of the individual flock and on local conditions. Consult a poultry pathologist, a poultry specialist, or your county agent.

It is desirable to have vaccinations completed at 16 weeks, or about 1 month before pullets are ready to lay. Under normal conditions, pullets should not be vaccinated as late as 20 weeks; this will delay egg production. Revaccinations are necessary if hens are to be kept for a second year of production.

Normally, a vaccine is most effective when it is administered alone. Stress from multiple vaccinations may cause severe losses. In spite of this, some poultrymen give simultaneous vaccinations for two diseases to save time and labor in handling birds.

Birds should be healthy and vigorous at the time they are vaccinated. Losses may be high if the flock is sick or has severe parasitic infestations. Do not vaccinate birds that have coccidiosis. If this disease appears, wait until the birds have recovered before giving them any type of vaccination.

The entire poultry flock should be vaccinated for the same disease at the same time. The vaccinated flock should be isolated from all other birds on the farm for 2 to 4 weeks.

Follow the manufacturer's instructions carefully. Store vaccine at proper temperatures, use fresh supplies, mix as directed, and administer exact dosages. If vaccines are misused, they may fail to protect the flock.

REARING STARTED PULLETS

A favorable environment for started pullets may be provided either in a poultry house or on range; healthy, vigorous replacement birds are grown under both systems.

The trend is to raise and to keep farm laying flocks entirely in confinement. Large pole-type houses are widely used for rearing birds.

Confinement takes less land and labor than range-rearing systems. Confined birds show fewer losses from coccidiosis, from worms and other parasites, and from such natural enemies as hawks, crows, and foxes. Although cannibalism often is a problem in confined flocks, it usually can be prevented or controlled by debeaking.

Range-grown pullets mature a few days later than confined birds. They sometimes have trouble adjusting to the limited space of the laying house.

Confinement

Birds may be kept in the same house in which they are brooded, or they may be transferred to a growing house when they are well feathered. Purchased started pullets usually are placed directly in a growing house.

Some extra space may be obtained by careful culling of the flock when birds are transferred to the growing house, or when cockerels are separated from the pullets, at about 6 weeks. Cull only obviously unhealthy or unthrifty birds from bred-to-lay stock.

Keep litter dry, and add new material as necessary. When roosts are not used, litter requires extra care.

Started pullets.

Parasites—especially lice—must be controlled. (See p. 35.)

Flocks with more than 2,000 chickens are easier to manage if the growing house is divided into pens. Put about 1,000 birds in each pen.

Perform the stress-producing jobs usually associated with moving pullets—vaccinations, debeaking, dusting for lice and mites—while birds are still in the growing house.

Install a few nests in the growing house before pullets start laying.

Range

Range-reared birds benefit from exercise, sunlight, and green feeds. Range may be used for pullets from the time they are brooded until they are ready for the laying house, or for cockerels to be used for breeding.

Birds on range may eat 5 to 15 percent less mash and grain than confined birds. Range rearing requires more labor than the rearing of confined birds; parasites and predators on range are major problems.

Often, it is difficult to find clean land for range. To avoid outbreaks of disease in range-reared chickens, clean range should always be used.

Clean range can be maintained by using a 2- or 3-year range-rotation program. At the end of a growing season after mature birds have been removed, the range is planted to pasture grass or cover crops. Droppings from the poultry house should not be used to fertilize the area.

An acre of good range can support 300 to 500 growing pullets. Allow the birds to run freely. Do not crowd the range, and do not mix birds of different ages in one area.

A range shelter usually is a small, portable building with wire walls and a wire floor. Doors should be high enough so an attendant can enter. Normally, shelters are placed at least 250 feet apart. They should be moved as often as necessary to prevent the range from becoming highly contaminated with droppings, or the sod from being killed.

The season will affect the time that birds are placed on range.

Before moving birds to range, clean and repair range shelters. When shelters are in place, screen the area under the house to keep pullets out of droppings. It often pays to move pullets at night.

Birds on range should have water available at all times. Drinking fountains should be placed within 15 feet of feeders and spaced evenly. Because birds ordinarily will not go more than 100 feet from the shelter to get water, there should be at least one waterer at each shelter. When shelters are moved, the water supply also should be moved. This will minimize wet spots on range.

Shade is another requirement for range. Natural shade from trees and shrubs should be used where it is available.

Water and feed should be placed in shade.

Mow the range frequently to keep grass tender and succulent.

Do not let droppings beneath the range shelter pile too high. Keep the droppings below the floor level of the shelter by removing or spreading them when necessary.

Remove droppings from range after moving shelters. Do not spread them on fields that will later be used for poultry, and do not pile them where pullets will have access to them.

To prevent losses from animals and wild birds, lock pullets in range shelters at night. The evening lockup and early morning release are essential to sound management.

Debeak, treat for parasites, and complete other stress-producing jobs while birds are on the range.

MANAGEMENT OF LAYERS

Egg-laying rates of the old hens and of the replacement pullets affect the time that the laying house should be emptied and prepared for the new flock.

When egg production drops to an unprofitable level at the end of the flock's first laying year, the old hens should be removed from the house. Ideally, this will coincide with the time that 10 percent of the young pullets have begun to lay.

Under normal circumstances, it is not economical for a farm-sized flock that produces market eggs to carry hens through the molt at the end of their first year. During the 2 to 4 months that hens are out of production, they eat 6 to 7 pounds of feed each month.

It may be profitable to carry high-producing layers through a molt and keep them for a second year if—
- Egg prices are high.
- Birds are healthy and vigorous.
- Separate housing is available.

If you maintain second-year birds in your flock, select layers with care. Keep only the best producers. Take special precautions to prevent disease and avoid stresses.

Preparing the House

To prepare the laying house for young pullets, thoroughly clean and thoroughly disinfect the interior. These steps usually prevent the carryover of diseases and parasites from the old to the new flock.

Allow the laying house to remain empty at least a week after it is cleaned and disinfected.

Check feeders and waterers. Make necessary repairs to the house and equipment.

Install nests and fresh nesting material, such as shavings, ground oystershells, rice or oat hulls, sawdust, or a commercial preparation. There is more trouble with floor eggs if nests are not put in the house before pullets are housed.

Put down new litter.

Moving Pullets

Before moving pullets to the laying house, examine them individually. Remove weak, runty, and obviously sick chickens from the flock. Cull pullets that appear to be unthrifty or extremely slow maturing; they are unlikely to be profitable layers.

Pullets are easily disturbed. Handle them gently. Do not make unnecessary noises or frighten them. Because pullets become less excited when they are handled in the dark, many poultrymen move pullet flocks into laying houses at night.

Care of Pullets

Watch pullets after they are placed in the laying house; do not let them pile up or smother. Close nests at night. A night light will help birds find roosts.

Clean nests regularly; add new nesting material as needed to produce clean eggs. Clean waterers at least once every day and feeders whenever they need it. Stir litter and add new material frequently; remove and replace wet and "caked-over" spots, as necessary. Control lice and other parasites.

Lighting

Layers need 14 hours of light per day throughout the year. If light is irregular, hens may go into a premature molt.

Artificial lights are used in houses with windows to supplement available natural light. To maintain 14 hours of daylight in the laying house, use artificial lights on a regular schedule as soon as total natural daylight drops below this level.

In windowless houses, all light is artificial.

Popular lighting plans include:

- All-artificial lighting. Lights are turned on 14 hours a day. Usually, an automatically controlled lighting clock is used.
- Morning lighting. Lights are turned on 2 to 5 hours before daybreak to provide a longer "working day" for hens. They are turned off at daybreak.
- Evening lighting. Lights are turned on at sundown to extend the hens' "working day." Before the end of 14 hours of light, dimmers give warning that lights will soon go out.
- Morning-evening lighting. Lights are turned on before sunrise and again after dark to provide a total of 14 hours of light. Dimmers operate before lights are turned off.
- All-night lighting. Lights are turned on at dusk and turned off at dawn. Low-wattage lamps, which provide just enough light for birds to see feed and water, are used. All-night lights are normally used for yearling hens.

Culling

Proper culling is essential for a profitable farm laying flock.

Culling involves identifying and removing unproductive birds. Such birds are an economic drain and can reduce the net profits of the flock significantly.

Do not overcull bred-to-lay birds. Flocks of these birds normally maintain high annual rates of egg production. There are relatively few culls unless the flock has been severely diseased or mismanaged.

Selection should begin as soon as birds arrive on the farm. Obviously deformed, weak, and diseased chickens should be removed as soon as they are observed. After the flock begins to lay at a profitable rate, nonlayers and low producers should be culled regularly.

In farm laying flocks, culling usually is based on one or a combination of these guides to egg production:

- Egg-laying indicators.
- Bleaching of pigment.
- Molting.

Guides to culling and additional information about identifying unproductive pullets are discussed in the USDA publication on culling hens.

High producer (left) and nonlayer (right).

Marketing Eggs

Markets available to the egg producer are largely dependent on the geographical location of the farm. Most farm eggs are originally sold in the immediate area in which they are produced. Produce dealers, cooperatives, shippers, and hucksters buy eggs at wholesale prices and in wholesale lots.

Poultrymen near cities may sell eggs directly to consumers on house-to-house routes or at roadside stands. Direct sales bring higher prices and—because handling costs are reduced—usually higher profits. However, if extra labor must be hired to process and sell eggs, retail operations may not pay.

High-Quality Eggs

Consumers want eggs with fresh-laid appearance, flavor, and nutritive value. The shell should be strong, regular, and clean; the white (or albumen), thick, clear, and firm; the yolk, light colored, well centered, and free from blood and meat spots.

Egg characteristics are inherited. Bred-to-lay birds have been produced through careful breeding programs that developed desirable egg traits as well as sustained laying ability.

Feed also affects egg quality.

Lack of calcium causes eggs to have soft or thin shells. Other reasons for poor shells are insufficient minerals, lack of vitamin D, high temperatures in the laying house, and diseases. Faulty shells also may be the result of inherited factors.

To produce infertile market eggs, keep males out of the laying flock.

Care and Handling of Eggs

Eggs produced clean and kept clean are most profitable.

Eggs are perishable. They deteriorate rapidly without proper care. Quality can best be preserved by keeping eggs cool and humid from the time they are laid until they are used.

As soon as possible after a market egg has been laid, it should be gathered and cooled.

Gather eggs frequently—at least three times a day. In warm weather, increase collections to four or five times a day. Frequent gathering reduces loss of interior quality from heat and increases the number of sound, clean eggs available for sale.

Handle eggs carefully. Keep hands and equipment clean.

Use open plastic or rubber-coated heavy wire baskets to gather and hold eggs. These permit free air circulation, so the eggs cool in about half the time that is required in closed containers.

Place baskets in an egg storage room to cool. Keep the temperature of the egg room between 45° and 55° F and the relative humidity at about 70 percent.

Check egg-room conditions and adjust cooling equipment when necessary. Temperatures above 65° F will cause eggs to deteriorate. Eggs freeze and shells often break at temperatures below 28°. If humidity is above 80 percent, eggs may become musty and moldy.

Use the egg room only for eggs. Keep eggs away from onions, kerosene, and other products with pungent odors that eggs might absorb.

If a high percentage of eggs are dirty, review your management practices. Spending time to produce

Air-conditioned egg room.

clean eggs is more profitable than cleaning dirty eggs.

Despite cleanliness and good management, some soiled eggs are produced.

The egg dealer usually specifies the method of cleaning to be used. Some require that soiled eggs be drycleaned; others, that eggs be washed on the farm. Handlers who have their own cleaning equipment often prefer that poultrymen do not wash the eggs on the farm.

Drycleaning usually is done by buffing dirty spots with fine sandpaper, steel wool, or emery cloth. Drycleaning is time consuming, and it may increase losses from cracks and crushed eggs.

Washing is the fastest way to clean a large number of soiled eggs. If eggs are to be washed on the farm, the egg-washing machine should be operated exactly as the manufacturer recommends.

If water temperatures or immersion times are not precisely controlled, or if eggs are washed improperly, the washed eggs will deteriorate more rapidly while they are being held. Therefore, it is considered good practice to sell washed eggs direct to consumers for immediate use.

Never wash clean eggs.

Use strong, clean, odorless containers to keep eggs in good condition during shipping and storage. Precool containers, fillers, and flats by placing them in the egg room about 24 hours before packing eggs.

Eggs for the wholesale market usually are packed in fiber, wood, or wire-bound cases that hold 30 dozen eggs. If secondhand cases are used, they must be clean and in good condition.

Market farm-produced eggs twice a week. Eggs should be kept at temperatures between 45° and 60° F when they are transported to market to maintain highest quality.

All eggs sold direct to the consumer should be candled to determine their interior quality. By turning the egg carefully, the candler can observe the condition and size of the air cell, yolk, and white. The light reveals defects, such as blood spots, blood rings, meat spots, and development of the germ spot in fertile eggs.

Eggs for retail sales may be packed in cartons containing two rows of six eggs each, or three rows of four eggs.

Pack all eggs with the large end up. The egg maintains its normal physical balance in this position. Properly packed eggs have fewer stuck yolks.

Graded eggs sell for higher prices than ungraded ones. Grades, which are based on standards established by the U.S. Department of Agriculture, indicate egg quality.

HATCHING-EGG PRODUCTION

Before beginning a hatching-egg flock, make a definite agreement with a hatchery. Reliable hatcheries with year-round incubation schedules set standards of breeding, sanitation, disease control, and egg handling for their supply flocks. Breeding flocks should meet the predetermined requirements of a hatchery or a classification of the National Poultry Improvement Plan.

If the existing farm flock is to be used for hatching-egg production, chickens must be blood-tested. All hatcheries participating in the Plan must use supply flocks that are negative to the test for pullorum disease and fowl typhoid. Tests are supervised by employees of the official State agency that administers the provisions of the Plan in cooperation with the U.S. Department of Agriculture. Reactors to pullorum disease and to typhoid are removed from the flock, and retests are required until no evidence of infection is found.

Many hatcheries provide male chicks of a special blood line to be grown on the farm. Some hatcheries supply both male and female chicks for the farm breeding flock. In this way, the hatchery can control the inheritance of the day-old chicks it later will offer for sale.

For a hatching-egg flock, start 10 to 14 cockerels for each 100 pullets; cull down to 8 or 12 when cockerels are 10 weeks old. At mating, provide 1 cockerel for each 15 to 17 Leghorn hens. Heavier breeds—Rhode Island Reds, Plymouth Rocks, and New Hampshires—require 1 cockerel for each 12 to 15 hens.

Birds in the breeding flock should conform to the characteristics desirable for their breed, variety, cross, or other combination of breeding. They should be developed for quality eggs, high production, and disease resistance.

The National Poultry Improvement Plan provides for the upgrading of participating flocks through breeding and disease control.

Chicks should be placed in a clean, well-regulated brooder as soon as they are brought to the farm. Follow the practices used to brood chicks for the laying flock (p. 19).

Breeding Flock Management

Traditionally, cockerel chicks have been separated from the pullets in the breeding flock at 6 weeks. However, an increasing number of poultrymen are raising cockerels with pullets. This reduces the fighting among males and the disturbance among females at breeding time.

Good housing and proper equipment, careful feeding, protection against extreme heat and cold, and other points of sound management contribute to production of a large number of high-quality, fertile hatching eggs.

Vaccination plans may call for revaccination of breeding stocks for

Male of White Cornish type in broiler breeding pen.

Newcastle disease and bronchitis at least 1 month before mating. Chicks from recently vaccinated hens obtain temporary immunity to these diseases.

If yearling hens are used, force them out of production to allow at least a 2-month rest before they begin laying hatching eggs.

To force a molt, turn off lights; feed grain instead of mash. If birds do not begin to molt, remove water for half days. When birds are molting, resume regular mash or pellet feedings. Introduce breeder mash when molting birds begin to gain weight.

Cull again before placing birds in breeding pens.

Mate birds at least 2 weeks before the hatchery wants its first hatching eggs. Provide 6 cockerels for each 100 Leghorn pullets, or 8 cockerels for each 100 heavy-breed pullets. Too many birds in a breeding pen will reduce matings.

Save several extra males out of each breeding flock to use for replacing cull or diseased cockerels.

Provide 14 hours of light per day for a pullet breeding flock. Keep lights and fixtures clean. Do not use artificial light during molting.

Care of Hatching Eggs

Gather hatching eggs three or four times a day. When temperatures are extremely hot or cold, gather eggs every hour. Handle hatching eggs like market eggs (p. 26). Adequate hatchability can be maintained if eggs are handled properly. Hatching eggs, as well as market eggs, should be clean.

As soon as possible after they are laid, store hatching eggs at a temperature between 50° and 60° F. Humidity should be 80 percent.

Select eggs for hatching that meet the weight requirements of your market, and are normal in size, shape, color, and shell texture. Excessively large or small eggs often are infertile; do not send them to the hatchery.

Before packing eggs, examine them individually. Remove those with obvious defects.

Pack hatching eggs large end up. Keep them cool. If possible, arrange for pickup or delivery twice a week. After eggs have been held on the farm 7 days, tip egg crate sharply to prevent yolk from sticking to shell; repeat once or twice daily. Do not hold hatching eggs on the farm any longer than necessary; never hold them more than 2 weeks.

MEAT PRODUCTION

Birds grown for meat—capons, roasters, and broilers—usually are crosses or hybrids. Cornish varieties—Silver and Dominant White—are popular for use as the male parent because their chicks develop meaty breasts and legs. Varieties and strains with white plumage often are used for female parents because they produce birds with excellent market appearance.

Hatcheries, breeders, and dealers are continually improving the stock they sell for meat production. Check the growth rate, feed conversion, and market acceptance records of available stocks before ordering meat-type chicks.

For meat production, almost all flocks are started from day-old chicks. Birds usually are purchased as straight-run chicks—about half pullets and half cockerels.

Housing

Broilers, capons, and roasters may be raised either in confinement or on range. The tendency is to raise them in confinement; pole-type houses are extensively used.

In a floor housing system, space requirements for all types of chickens are the same until birds are 10 weeks old. Provide ½ square foot per bird until chicks are 2 weeks old and 1 square foot per bird between 2 and 10 weeks. For capons and roasters from 10 to 20 weeks old, allow 2 to 3 square feet of floor space per bird. When meat-type birds are over 20 weeks old, provide 4 to 5 square feet per bird.

Most broiler growers provide lights 24 hours a day as long as birds are on the farm. A usual lighting system allows one 60-watt lamp for each 200 square feet of floor space, or about ¼ watt per square foot.

For each 100 meat-type birds, provide the same kind and amount of equipment required for 100 pullets. However, no nests or roosts are needed.

Brooding

Capons, roasters, or broilers should be brooded the same way as replacement chicks.

Advance preparations, sanitation, and management practices used for raising replacement chicks should be followed for raising capons, roasters, and broilers on the farm. (See pp. 14 and 19.)

To avoid cannibalism, day-old meat-type chicks often are debeaked at the hatchery.

Follow the vaccination schedule recommended by a poultry specialist or a poultry pathologist.

Caponizing

A capon is any male chicken that has been castrated. Following this surgical operation, the bird fattens more readily, and produces more tender meat. Capons sell for a higher price per pound than broilers or roasters, because they require more labor and cost more to produce. They do not use feed as efficiently as light-weight birds.

Most poultrymen caponize when cockerels are 3 to 5 weeks old. The operation is of little value after chickens are 2 months old.

After the operation, put capons in clean quarters, separated from other chickens for 2 or 3 days. Capons do not need special care after they recover from the initial shock of the operation.

For directions on how to caponize, consult the U.S. Department of Agriculture publication or see your county agent or a poultry specialist.

A compound, dienestrol diacetate, sometimes is added to special feeds to produce a feminizing effect on male chickens. This may result in fattening and tenderizing of rapidly growing birds. If you use this compound, follow the directions of the manufacturer.

TABLE 3.—*Complete broiler feeds*

Ingredient [1]	Starter	Finisher
	Percent	*Percent*
Ground yellow corn	59.40	66.61
Fish meal (60 percent)	6.00	5.00
Poultry byproduct meal	5.00	2.50
Corn gluten meal	4.00	
Soybean oil meal (44 percent protein)	18.00	16.50
Dried whey		1.50
Alfalfa meal (17 percent protein) [2]	2.00	2.00
Dried distillers' solubles	2.50	2.00
Calcium carbonate	1.25	1.00
Dicalcium phosphate (18 percent phosphorus)	.60	
Bonemeal, steamed		1.50
Salt, iodized	.30	.30
Manganese sulfate (65 percent grade)	.05	.05
Vitamin A supplement (4,000 USP units per gram)	.05	.05
Vitamin D_2 supplement (1,500 ICU per gram)	.06	.06
Vitamin B_{12} supplement (12 milligrams per pound)	.05	.05
Riboflavin supplement (227 milligrams per pound)	.50	.50
Choline supplement (25 percent grade) [3]	.10	.10
DL-methionine (feed grade)	.04	.18
Antibiotic supplement (10 grams per pound) [4]	.05	.05
Arsonic acid (10 percent) [5]	.05	.05
Total	100.00	100.00
	Grams per ton	*Grams per ton*
Niacin	25	25
Calcium pantothenate	5	5
Vitamin E	5	5
Vitamin K	1	1

[1] Should also include a coccidiostat at level recommended by manufacturer.
[2] Meal with guaranteed vitamin A content of 100,000 International Units per pound.
[3] Contains 25 percent choline chloride.
[4] Follow the recommendations of the manufacturer.
[5] Contains 10 percent 3-nitro, 4-hydroxyphenylarsonic acid. Other compounds, including sodium arsanilate or arsanilic acid, may be used at a level recommended by manufacturer. All arsenical compounds to be used in poultry feeds are subject to approval by Food and Drug Administration, U.S. Department of Health, Education, and Welfare.

Feeding Meat-Type Birds

Broilers need a feed that contains 20 to 24 percent of protein for the first 6 weeks. The exact percentage of protein needed depends on the feed's energy content.

All broiler rations should be fed without grain. A sample starter ration for broilers is listed in table 3 (above). Chick-sized granite grit may be fed in boxes or scattered on top of mash.

When using commercial feed, follow the manufacturer's directions.

At about 6 weeks, place broilers on a finishing mash that has an increased energy level and reduced protein level. A sample finishing mash is listed in table 3. Pellets often are used for broilers from 6 weeks until they reach market weight.

Give capons and roasters the kind and amount of feed recommended for broilers during the first 6 weeks. After changing to finishing mash, supply cracked corn to roasters and capons in the afternoon. Gradually increase the grain until birds are getting equal amounts of corn, mash, and pellets at 12 weeks of age.

For roasters, increase the corn in the diet to 50 percent after the 15th week. Continue to supply grit as long as birds are fed whole grain.

Management

Commercial growers usually raise at least four flocks of broilers a year, and market about 97 percent of the birds they start. They produce a pound of meat on about 2.4 pounds of feed. To compete with this market, the farm broiler flock must be raised efficiently.

Some farm broiler producers follow a four-times-a-year brooding schedule. Others grow one or two flocks to make use of housing that otherwise would be empty. If straight-run medium-weight chicks are ordered for the replacement flock, cockerels may be raised as broilers.

Broilers, roasters, and capons—whether raised in confinement or on range—require the same care as replacement chicks.

Use good management practices:
- Start with quality chicks.
- Clean the quarters before housing birds. Keep litter clean and dry.
- Brood birds carefully.
- Provide enough space (about 0.8 to 1 square foot per bird).
- Provide adequate feed and water.
- Vaccine birds if necessary.
- Provide adequate light. In hot weather, use all-night lights to encourage eating.
- Do not mix birds of different ages.
- Debeak birds when necessary.
- Watch for diseases and parasites; control those that do appear.
- Remove diseased birds from the flock. Promptly burn or dispose of birds that die of disease.
- If range is used, provide clean ground with adequate green feed.

Marketing Poultry

Poultry may be sold live, dressed, or ready-to-cook.

Before deciding whether to sell poultry live or to process it on the farm, consider the markets available and price variations. Live birds may be sold at the farm, or delivered to an auction house, dealer, broker, or processing plant. Farm-processed poultry may be sold on the wholesale or retail market.

Top-quality dressed or ready-to-cook birds can be obtained only from live birds that are in prime condition at the time of slaughter. Do not feed birds for 12 hours before slaughter. During this period, give them plenty of water, which will flush out feed from the crop and intestine.

Handle live birds carefully to minimize losses from bruises, smothering, and shrinkage. Deliver them in the early morning, if possible. Do not crowd birds in delivery coops.

DISEASES, PARASITES, AND PESTS

Prevention is the most satisfactory way to deal with poultry diseases and parasites. Many disease conditions can be prevented through good management, but cannot be cured once they actually occur.

Contagious diseases usually can be prevented by completely isolating the flock; vaccinations are useful against some common diseases. (See p. 21.) Many inherited diseases can be avoided by using tested, healthy stock. Diseases caused by nutritional deficiencies seldom develop if chickens are fed balanced rations.

Some drugs are effective in preventing specific diseases or parasites, but they do little good when used haphazardly. Drugs should be used to reinforce good management, not as a substitute for it.

The following are effective preventive measures:

- Clean the poultry house before moving the flock in.
- Provide clean, fresh feed and water. Treatment of water is not necessary unless the source is contaminated.
- Keep poultry house and equipment clean and dry.
- Provide proper ventilation.
- Screen houses to protect the flock from diseases carried by free-flying birds.
- Control rats and mice.
- Obtain U.S. Pullorum-Typhoid clean chicks from a National Plan hatchery.
- If possible, have birds of only one age on the farm. Alternatively, separate birds of different ages into different flocks. Keep flocks at least 40 feet apart.
- Keep poultry house locked.
- Limit visitors to the poultry house. When a visit is justified, supply the visitor with rubber or plastic boots or insist that he scrub and disinfect his shoes or boots.
- Store droppings and litter where they are not accessible to range or wild birds.
- Use a vaccination program recommended for your area.
- Keep delivery trucks as far away from the flock as possible. Disinfect truck tires before delivering feed to range.
- Do not let trash or junk accumulate in or near the poultry house.

Disease Control

When a disease outbreak occurs, determine the cause as soon as possible. If you do not recognize the disease or parasite, take or send four or five live chickens to the nearest poultry diagnostic laboratory. (Your veterinarian or county agent can give you the address.) Poultry-disease experts need live birds to make an accurate diagnosis. In sending chickens to the laboratory, be sure to list the symptoms, number of affected birds, number of deaths, source of stock, size of flock, feeding program, vaccines used, your name, address, and county, and other pertinent information.

Recommended measures to control disease:

- Incinerate or put into a disposal pit all dead chickens.
- Remove all obviously sick chickens from the flock. It usually is

Disposal pit. Good sanitation requires immediate disposal or incineration of dead birds.

best to kill them and dispose of the carcasses. If birds are to be treated, put them in a separate pen or hospital coop as far away from the other birds as convenient. Follow the treatment suggested by the poultry laboratory or a poultry pathologist.

• Be careful not to spread disease from the sick birds. Take care of the diseased birds last; wear clothing and boots that can be left outside the pen, or disinfected after use. Wash your hands before working with other chickens.

• Clean and disinfect the poultry house as thoroughly as possible after diseased birds are removed, without disturbing the remaining flock. Remove litter used by diseased birds and replace it. Clean and disinfect equipment.

• To get rid of internal parasites, contact your county agent or a poultry specialist.

Mites, Lice, and Ticks

Mites, lice, and ticks are common external parasites.

If lice and mites occur, use one of the following methods of control:

• Paint roosts with a 3-percent malathion emulsion, or 1-percent Rabon emulsion.

• Treat floor litter with one of the following: 4- or 5-percent malathion dust, 3-percent Rabon dust, or 0.5-percent coumaphos dust.

• Spray floors, roosts, and walls with 1-percent malathion spray, 0.25-percent coumaphos spray or 0.3-percent naled.

• Spray chickens with 0.25-percent coumaphos spray, 0.5-percent Rabon spray, or 0.5-percent malathion spray.

• Treat chickens individually with a 4-percent malathion dust, a 0.5-percent coumaphos dust, or a 1-percent Rabon dust.

If fowl ticks occur, apply a spray containing 3-percent malathion or 1-percent Rabon to roosts and to the interior of poultry houses.

Read and follow the directions on the label.

Rats and Mice

Rats and mice kill young chickens, destroy eggs, eat or contaminate poultry feed, and damage buildings by gnawing wooden walls, foundations, and equipment. They also spread many diseases and parasites.

Losses from rats and mice in a farm-sized flock may total several hundred dollars a year, including chickens, feed, and other destruction. The damage often goes unnoticed, because losses are gradual and because rats and mice seldom appear when a man is in the poultry house.

To make new buildings ratproof, use concrete foundations and floors. Metal shields around doors and screens over small openings are effective in keeping rodents out.

In fighting rats and mice, get rid of their breeding and hiding places on the farm. Clean out trash, dumps, piles of old lumber or manure, and garbage. Find and block runs or burrows. Then use a poison.

You will find information on rat poisons and how to use them in Wildlife Leaflet 402, "Anticoagulant Rodenticides for Control of Rats and Mice." Send your request on a post card to: Publications Unit, Bureau of Sport Fisheries and Wildlife, Fish and Wildlife Service, U.S. Department of the Interior 20240. Be sure to include your ZIP code number in your return address.

If you have further questions, consult your county agricultural agent.

Predators

Confinement rearing has greatly reduced poultry losses from hawks, owls, crows, foxes, skunks, weasels, and other predatory animals. These wild animals and birds often attack chickens on range. Young chickens are especially vulnerable.

To protect birds on range from predators, close shelters at night. Screen the space beneath the shelters. Install an electric wire slightly above ground level on the outside of the fence that encloses the range. Use steel jump traps on top of fence posts to catch crows, owls and hawks.

Sometimes, shooting is the most effective way to get rid of wild animals and birds. Before trapping or shooting animals and birds, find out about local and State laws from wildlife authorities, law enforcement officers, or your county agent.

PRECAUTIONS

Federal and State regulations require registration numbers on all pesticide containers. Use only pesticides that carry this designation. Read and follow all directions on the label.

USDA publications that contain suggestions for the use of pesticides are normally revised at 2-year intervals. If your copy is more than 2 years old, contact your Cooperative State Extension Service to determine the latest pesticide recommendations.

The pesticides mentioned in this publication were federally registered for the use indicated as of the issue date of this publication. Because the registration of a pesticide that you have had in your possession for some time can be changed, you may wish to check with your local agricultural authorities to determine the registration status of the pesticide.

Additional Information About Poultry

The U.S. Department of Agriculture has issued a number of publications on poultry and related subjects.

To obtain copies of the following publications, ask your county agricultural agent or write to the Office of Communication, U.S. Department of Agriculture, Washington, D.C. 20250. Include ZIP Code with your return address.

	Order No.
The Fowl Tick: How To Control It	L 382
The House Fly: How To Control It	L 390
Cecal Coccidiosis of Chickens: How To Control It	L 393
Lighting Poultry Houses	FB 2229
Houses and Equipment for Laying Hens	MP 728

The publications below may be obtained by writing to National Poultry Improvement Plan, Beltsville, Md. 20705. Include ZIP Code with your return address.

National Poultry Improvement Plans and Auxiliary Provisions	ARS-NE-32-1
Participants in the National Poultry Improvement Plan	ARS-NE-9-2
Report of Random Sample Egg-Production Tests	ARS-NE-21-1

For further information about farm poultry flocks, consult—
- Your county agricultural agent.
- Your State Extension Service poultry specialist.
- The poultry staff of your State agricultural college.
- Your State poultry disease diagnostic laboratory.
- Service men for contract flocks.
- Feed dealers.
- Your State Department of Agriculture.
- The U.S. Department of Agriculture, Washington, D.C. 20250.
- Weekly and monthly periodicals about poultry.

www.ingramcontent.com/pod-product-compliance
Lightning Source LLC
Chambersburg PA
CBHW081122240526
45470CB00019B/2922